John Lewis-Stempel is a writer and farmer. His books include the *Sunday Times* bestsellers *Woodston, The Running Hare* and *The Wood*. He is the only person to have won the Wainwright Prize for Nature Writing twice, with *Meadowland* and *Where Poppies Blow*. In 2016 he was Magazine Columnist of the Year for his column in *Country Life*. He farms cattle, sheep, pigs and poultry. Traditionally.

www.penguin.co.uk

The
Soaring Life
of the
Lark

John Lewis-Stempel

doubleday

TRANSWORLD PUBLISHERS
Penguin Random House, One Embassy Gardens,
8 Viaduct Gardens, London SW11 7BW
www.penguin.co.uk

Transworld is part of the Penguin Random House group of companies
whose addresses can be found at global.penguinrandomhouse.com

First published in Great Britain in 2021 by Doubleday
an imprint of Transworld Publishers

A CIP catalogue record for this book
is available from the British Library.

ISBN 9780857525802

Typeset in 11/14.5pt Goudy Oldstyle Std by Jouve (UK), Milton Keynes
Printed and bound in Great Britain by Clays Ltd, Elcograf S.p.A.

The authorized representative in the EEA is Penguin Random House Ireland,
Morrison Chambers, 32 Nassau Street, Dublin D02 YH68.

Penguin Random House is committed to a sustainable future
for our business, our readers and our planet. This book is made
from Forest Stewardship Council® certified paper.

CONTENTS

The Skylark

The rolls and harrows lie at rest beside
The battered road; and spreading far and wide
Above the russet clods, the corn is seen
Sprouting its spiry points of tender green,
Where squats the hare, to terrors wide awake,
Like some brown clod the harrows failed to break.
Opening their golden caskets to the sun,
The buttercups make schoolboys eager run,
To see who shall be first to pluck the prize –
Up from their hurry, see, the skylark flies,
And o'er her half-formed nest, with happy wings
Winnows the air, till in the cloud she sings,
Then hangs a dust-spot in the sunny skies,
And drops, and drops, till in her nest she lies,
Which they unheeded passed – not dreaming then
That birds which flew so high would drop again
To nests upon the ground, which anything
May come at to destroy. Had they the wing
Like such a bird, themselves would be too proud,
And build on nothing but a passing cloud!
As free from danger as the heavens are free

From pain and toil, there would they build and be,
And sail about the world to scenes unheard
Of and unseen – Oh, were they but a bird!
So think they, while they listen to its song,
And smile and fancy and so pass along;
While its low nest, moist with the dews of morn,
Lies safely, with the leveret, in the corn.

John Clare

PROLOGUE

L ONG AGO, THE Venerable Bede calculated that the first day of Creation was the eighteenth of March.

Remembrance of this schoolboy fact comes to me as I stand in Bank Field, the dew on the grass as shiny as birth fluid, every hedge exhilarated by thorn's white blossom, the sky so clear I can see into Heaven. The astronomical arithmetic that led to the Anglo-Saxon monk's discovery was explained to me at my Form Three desk, agilely so. I thought the more plausible explanation was that Bede merely stood in an English field in March at dawn.

The world feels brand new on a spring morning such as this.

Certainly, the skylarks have found something to sing about. Two are fluttering above the hayfield, up into the dome of the blue world.

I was not up with the lark; the dawn-time. This is twenty minutes past dawn, although black night is still trapped down in the folds of the mountains of Wales to the west.

In their joy for life, skylarks will sometimes climb the sky and sing before first light. Once, when I was a child, I walked through our village with my parents at 3 a.m., after a wedding party. We took the footpath through the meadows, and every dark one of them fizzed with skylarks rising. But that was in the birdy 1970s.

How to describe the ecstatic song of larks? How the writers and poets have tried. The Victorian naturalist Richard Jefferies settled for a 'waterfall in the sky' as he remembered sitting on the South Downs in his autobiographical *The Story of My Heart*; and for 'sunshine in song' in 'Out of Doors in February'. Percy Bysshe Shelley famously plumped, in 'To a Skylark', the most celebrated of skylark poems, for 'flood of rapture' to describe the melodious chattering of the 'blithe spirit'. Better to my ear this morning is the stanza of Scotsman William Dunbar, nearly as ancient as Bede:

> *Through beamis red, gleming, as ruby sparks,*
> *The skyes rang for the shouting of larks.*

But best is George Meredith in his 'The Lark Ascending', whose skylark:

> *. . . rises and begins to round,*
> *He drops the silver chain of sound*
> *Of many links without a break,*
> *In chirrup, whistle, slur and shake,*
> *All intervolv'd and spreading wide,*
> *Like water-dimples down a tide*
> *Where ripple ripple overcurls*
> *And eddy into eddy whirls;*
> *A press of hurried notes that run*
> *So fleet they scarce are more than one . . .*

Meredith's poem, of course, inspired Ralph Vaughan Williams to write his musical work of the same name, with quivering violin imitating the ascension singing of *Alauda arvensis*. I first heard Vaughan Williams's *The Lark Ascending* on a cassette tape borrowed from Hereford City Library when I was eleven; I returned the tape, but the music plays still in my mind.

Written as Vaughan Williams walked Margate's cliff top in the first week of the Great War, it was an attempt to capture the spirit of English pastoralism. He succeeded.

We poeticize and musicalize the bird; the French, as the deceptively jaunty song '*Alouette*' confirms, roast it. ('Little skylark, I'll pluck your feathers off' runs the refrain, translated.) Yet we also do harm to the skylark; the numbers of the bird in England plummeted by 24 per cent between 1995 and 2013 due to changes in 'farming practice', principally the multiple cutting of grass for silage – pickled winter fodder for livestock. Skylarks are doughty little birds but they cannot, sitting in their nest on the ground, survive the swirling steel blades of a Claas Disco mower three, four times a year. The big switch from autumn to spring sowing of cereal crops has done them no good either, because the crop lacks the right cover at the right time. I suppose, in retrospect, that the male skylarks of the 1970s needed to sing earlier in order to rise above the competition that came from numbers. Or maybe, in those days of less intensive agriculture, they just had more to sing about. Too many fields today, where skylarks once sang, are silent.

I wrote a book called *The Running Hare*, about the resistible decline of wildlife on farmland. I could just as easily have called the book *The Descending Skylark*. The hare and the skylark: they are both iconic animals, and indicators of all that is well or unwell in the countryside.

I watch the skylarks over the hayfield become trembling specks, then dissolve in the divine heights. A skylark can ascend 200 metres.

My love of songbirds is very British, very old. The

4

Anglo-Saxons of the ninth century preferred to name members of the avian suborder *Passeri* – which, strictly, biologically, is what a songbird is – not by their appearance, the preference of modern science, but by their song. An Old English calendar for 1061 has 364 days devoted wholly to the doings of saints. The remaining day? The eleventh of February, for which the entry reads 'At this time, the birds begin to sing.'

A spoilsport contemporary scientist will doubtless pipe that birds really only sing for a spouse and a tract of land. Maybe it is the fancy of frail humans to believe the birds sing for us. But listen to a skylark's song; surely it is Nature's own music?

In the gently falling shower of song from the unseen birds, I turn to my jobs, of checking sheep, of checking cows on a Herefordshire hill farm. There is growth in the grass. It is spring. And the skylarks are its heralds.

The Skylark

The earth was green, the sky was blue:
I saw and heard one sunny morn,
A skylark hang between the two,
A singing speck above the corn;

A stage below, in gay accord,
White butterflies danced on the wing,
And still the singing skylark soared,
And silent sank and soared to sing.

The cornfield stretched a tender green
To right and left beside my walks;
I knew he had a nest unseen
Somewhere among the million stalks:

And as I paused to hear his song,
While swift the sunny moments slid,
Perhaps his mate sat listening long,
And listened longer than I did.

Christina Georgina Rossetti

THE·LARK AND ·HER YOUNG·ONES.

I

The Natural Life
of the Lark

'THE LARK', OF course, is the skylark, *Alauda arven-sis*; not its less famous cousin the woodlark, or the winter visitor, the shorelark.

The skylark has a split personality. Celebrated for its aerial singing, it can also be determinedly earth-bound, feeding and nesting on land, hence its folk name 'ground lark'. Or in Sussex, 'clodhopper'.

Skylarks are birds of open places, downs, dunes, hills, city wasteland. So, in Britain, they primarily inhabit farmland, since this provides the largest area of suitable habitat. It was the cutting down of the wildwood for agriculture 10,000 years ago that gave the skylark its advantage. About 71 per cent of the surface of Britain is farmland. The sound of the sky-lark has been the soundtrack of the British countryside for as long as humans have been here in numbers.

In the furrow and tussock where the skylark makes its home, it runs in rapid movements, rather than hopping. It is this movement which is most likely to catch the eye, of birdwatcher and predator alike, since the bird's streaky brown plumage perfectly blends into the farmland floor. Underparts are white, as is the rather fetching eyestripe. A further aid to physical identification is that both male and female have a clump of feathers on the crown erected into a crest when the bird is alert. The skylark, starling-sized at 17cm, weighing between 25g and 50g (depending on sex, and time of year), has white outer feathers in its tail which show when it flies off, and which it displays in courtship. These give it the country name of 'lintie white'. Travel flight is the characteristic undulation of the lark family.

The wings of male skylarks are around 5–10 per cent longer than those of females. Bird-dealers in Victorian England were well aware of this: male skylarks were songsters to be caged for entertainment, the female of the species were sold for meat. In *British Birds with Their Nests and Eggs* (1896–8), Arthur Butler described how bird-dealers:

> *take the bird in the left hand with the tail towards them, and with the right hand draw down the wing until the*

*point of the first long primary touches the tip of the outer-
most tail feather: the wing of the male being distinctly
longer than that of the female, the so-called 'shoulder'
then appears to be much more angular in the former than
in the latter sex. I have seen considerable numbers of
birds thus tested, the males being caged and the females
returned to the catchers, and I never knew the test to
fail: but females are rarely forwarded by experienced
bird-catchers, most of them being killed at the nets and
sold to the poulterers.*

Historically, farmers were prepared to aid and
abet the bird-catchers; the skylark eats the emerging
shoots in the corn field; the bulk of the bird's menu,
however, consists of small insects and seeds. Making it
in truth the farmer's friend.

During autumn and winter's cold, larks gather in
travelling flocks and occupy their time principally in
searching for food on the ground. If disturbed, they
rise in a scattered manner, wheel in the air, re-form a
flock, and fly to a new, nearby feeding area. Contemporary winter flocks typically number ten to twenty.
Overall, the UK's resident population of the skylark
is 200,000 pairs, though some leave to be replaced
by continentals; the reverse occurs in spring. Before
the twentieth century, the skylark migrations across

Europe were natural wonders, blackening the sky in their passing numbers. It is worth quoting the account of skylark migration of the Victorian ornithologist Henry Seebohm, for the sense of the spectacle, for the understanding of what we have lost. Seebohm was on Heligoland, an island in the North Sea off the coast of Germany, in October 1876:

> The night was almost pitch-dark, but the town was all astir. In every street men with large lanterns and a sort of angler's landing-net were making for the lighthouse. As I crossed the potatoe-fields [sic] birds were continually getting up at my feet. When I arrived at the lighthouse, an intensely interesting sight presented itself.

The whole of the zone of light within range of the mirrors was alive with birds coming and going. Nothing else was visible in the darkness of the night but the lantern of the lighthouse vignetted in a drifting sea of birds. From the darkness in the east, clouds of birds were continually emerging in an uninterrupted stream; a few swerved from their course, fluttered for a moment as if dazzled by the light, and then gradually vanished with the rest in the western gloom. Occasionally a bird wheeled round the lighthouse and then passed on, and sometimes one fluttered against the glass like a moth against a lamp, tried to perch on the wire netting, and was caught by the lighthouse man. I should be afraid to hazard a guess at the hundreds of thousands that must have passed in a couple of hours; but the stray birds which the lighthouse man succeeded in securing amounted to nearly 300. The scene from the balcony of the lighthouse was equally interesting; in every direction birds could be seen flying like a swarm of bees, and every few seconds one flew against the glass. All the birds seemed to be flying up wind, and it was only on the lee-side of the light that any birds were caught. They were nearly all Sky-Larks; but in the heap captured there was one Redstart and one Reed-Bunting. The air was filled with the warbling cry of the Larks; now and then a Thrush was heard; and once a Heron screamed as it passed by. The night was starless

13

and the town was invisible, but the island looked like the suburbs of a gas-lit city, being sprinkled over with brilliant lanterns. Many of the Sky-Larks alighted on the ground to rest, and allowed the Heligolanders to pass their nets over them. About three o'clock in the morning a heavy thunderstorm came on, with deluges of rain; a few breaks in the clouds revealed the stars, and the migration came to an end, or continued above the range of our vision.

The arrival of migrating skylarks to the shores of Britain was no less dramatic. One ornithologist wrote:

The rush of Sky-Larks that land on our eastern coasts in autumn is almost past belief. Towards the end of October, or during the first week in November, the number that pass over these marshes is enormous. When the migration is on, you may see the great army of birds passing at a moderate height for days together; and you may know that the grand flight is still continued during the night, by constantly hearing the little travellers calling to each other as they pass on, coming from over the sea, to the inland pastures. In the daytime many of the birds break out into song the moment they are over dry land, as if full of joy at making the passage safely.

There are two collective nouns for skylarks, an 'exaltation' and a 'bevy'.

The skylark's nest on the ground is described exactly in 'The Lark's Nest', a poem by John Clare.

> *Behind a clod how snug the nest*
> *Is in a horse's footing fixed!*
> *Of twitch and stubbles roughly dressed,*
> *With roots and horsehair intermixed.*

John Clare (1793–1864), the 'Northamptonshire peasant poet', knew the skylark well, wrote about it often and not just as poetic metaphor. He was the surest observer of the bird and its habits before the professional ornithologists of the twenty-first century.

The skylark is one of the very few British songbirds to nest on open ground. In the absence of a natural hollow, the hen bird will scrape a shallow bowl in the earth, which she then lines with grass, woven into a mat with the bill. Her greenish-grey and cryptically speckled eggs are laid at the rate of one per day.

Due to high predation rates, skylarks need to produce up to three broods a season. The standard clutch size for the British skylark is four eggs, but can vary

between two and six. The thirteen days of incubation are undertaken by the female alone. April is the earliest month for eggs; the last sitting is August. According to the Victorian naturalist Reverend C. A. Johns, the skylark 'displays great attachment to its young, and has been known, when disturbed by mowers, to build a dome over its nests, as a substitute for the natural shelter afforded by grass . . . and to remove its young in its claws to another place of concealment'.

Both male and female are committed parents. Alas, they are not dome-makers as suggested by the good Reverend, or carriers of fledglings. Chicks are entirely dependent on insects until fledging, favouring sawfly larvae, beetles, ants, spiders and grasshoppers. Skylark pairs stay together for the breeding season, and may even remain partnered for the entirety of their short lives.

*

No one would accuse the skylark of being handsome. It is a pleasant, friendly, perky-looking little bird. The beauty is all in the song.

The aerial song by the males is, biologically, a broadcast message for sex and territory. Combining two physically demanding tasks, vertical ascent and carolling, the song-flight ritual is one of the most demanding in animaldom. Skylarks can sing for half an hour, though four to five minutes is more usual. The bird goes higher in fair weather than foul.

Having reached the apex of its song-flight, the bird descends, not at a steady rate, but by 'parachuting' to invisible floors, where it hovers. According to William Yarrell (1861), 'an ear well tuned to his song can even then determine by the notes whether the bird is still ascending, remaining stationary or on the descent'. This tonal difference between the up and down flights gave birth to old rural fable that on its ascending flight the skylark is happy at the thought of entering Heaven; having been denied entry by Saint Peter, the lark's tune takes on a minor, melancholy key as it comes down. Although the skylark's song appears to our ears to be continuous, analyses show it to be made up of a series of discrete phrases or syllables punctuated by short gaps of between 30 and 100 milliseconds. The repertoire of individual skylarks varies from around 160 to over 460 syllables, and this complexity is only increased by the skylark's insertion into its chorus the mimicked notes of other birds. The birds most sampled by larks are waders such as curlew, lapwing and redshank.

Song is given mainly by the males, with around 95 per cent of all song being delivered within 100 metres of the nest. However, females have a subsong resembling a short and muted version of the male's that is most commonly heard during pair formation, mating and nest building, and as an anxiety call when predators approach the nest. Skylark song, which is usually delivered from fifty metres plus above ground, does more than flauntingly seek a mate and defend a territory; it sends a sound signal-warning to predators. A high-flying, loud-singing skylark is twice as likely to survive a merlin falcon's attack than a weak singer of a skylark. Indeed, merlins as they grow older and wiser avoid the divo skylarks.

When the performer skylark nears the earth, it stops its song and makes a short horizontal flight into herbage, disappearing as convincingly here as it earlier vanished into the sky above. Singing occurs all year round, with a peak in April, a definite diminution over summer to a redux in October, and hesitant half-heartedness through winter. The skylark will sing from a perch – it does so in captivity – but such song is slow and muted, more communication call than celebration of existence.

Morphology explains something of the skylark's singing style. Larks differ from all other songbirds in the structure of the syrinx, the bony-ringed resonating chamber at the lower end of the windpipe; the syrinx of larks lacks a pessulus, the cartilaginous or bony bar at the junction of the bronchi. The absence of a pessulus enables the bird's vibrato style of singing.

Larks on the wing are frequently the prey of the hobby, sparrowhawk, peregrine, the harrier, and the aforementioned merlin. In some regions skylarks account for a quarter of the merlin's diet. The principal predators of the skylark on the nest and ground are: hedgehog, weasel, stoat, brown rat, badger, fox, corvid, kestrel, harrier, grass snake, squirrel.

The skylark who survives predation may live, in the wild, typically for two years.

Cecil Smith in *Birds of Somersetshire* (1869) recalls a caged skylark that reached twenty years, 'and its song continued almost to the last'.

The British Larks

There are ninety-three species in the avian family *Alaudidae*. On the British List of Birds, British Ornithologists' Union, 2017, there appears: woodlark, white-winged lark, calandra lark, short-toed lark, lesser short-toed lark, bimaculated lark, skylark, black lark, crested lark. Most are rare birds indeed, lost travellers from afar (there has only ever been one sighting of the lesser short-toed lark, in Dorset in 1992). Aside from the skylark, only the woodlark is resident, and only the shore lark anything like a familiar visitor.

The Woodlark (*Lullula arborea*)

The shy relative of the skylark, and a considerably rarer bird, now restricted to a few south and south-west counties, where it inhabits the edge of deciduous woods and heathland. Slightly smaller than the skylark, similarly stripey-brown in coloration, also with a crown-crest of feathers. The diagnostics are the short tail without white edges, but with a black spot on the top of the wing. Even by lark standards, the flight is undulating, approaching a switchback ride in the air. It perches on any handy prominent perpendicular feature, from tree to telephone post. Song is like the skylark's but interjected with a liquid 'lu-lu-lu-' sound, hence its French name '*Alouette lulu*'. The bird also likes to sit in a tree to trill, but fleetingly so; the song always seems to be around a bend, or behind a bower. In the nest, built by the female in a depression in the ground among vegetation, are laid three to five greyish-white, brown-flecked speckled eggs. Two, sometimes three, broods are reared per annum. The woodlark is a notably early nester; the first clutch is in the nest by the end of March.

Gerard Manley Hopkins gamely tried to transliterate the bird's song into words in 'The Woodlark':

TEEVO cheevo cheevio chee:
O where, what can tháat be?
Weedio-weedio: there again!
So tiny a trickle of sóng-strain;
And all round not to be found
For brier, bough, furrow, or gréen ground
Before or behind or far or at hand
Either left either right
Anywhere in the súnlight.
Well, after all! Ah but hark—
'I am the little wóodlark . . .

Heretically, there are some who prefer the wood-lark's music to that of the skylark. The French composer Olivier Messiaen includes the fugitive sound of the woodlark in the *Catalogue d'oiseaux* (1956–8), where a woodlark briefly duets with a nightingale.

The Shore Lark (*Eremophilia alpestris*)

Winter visitor to (primarily) eastern England, where it remains true to its name, keeping to the strandline, with the occasional foray to nearby fields. Diet is invertebrates, crustacea and seeds. Off-season dress is drab, though the black and sandy-yellow facial markings can still be determined. In its breeding pomp back in the

northern tundra it sports the diabolic double crest which gives it one of its alternative names, 'Horned Lark'.

Like the skylark and woodlark, it has been immortalized in music, featuring in *Cantus Arcticus*, op. 61, an orchestral composition written by the Finnish composer Einojuhani Rautavaara in 1972. The second movement, 'Melancholy', features a slowed-down recording of the song of the shore lark.

The Naming of the Lark

What's in a bird name? Essence. The ancients noted the skylark – to the eye, nothing but a small drab thing – for its song-flight. The scientific name of the skylark, indeed the whole genus *Alauda*, probably comes from the Latin corruption of the Celtic 'great singer' (*al* = great, *awd* = song).

What's in a multiplicity of names for a bird? The gauge of human interest, the register of familiarity. The number of popular names a bird enjoys is a sure marker of the degree to which it impinges on consciousness. The skylark possesses a richness of country and folk names. Primarily, local nomenclature is descriptive, for instance 'sky flapper' (Somerset) and 'rising lark' (Northamptonshire). Then there are names which are a variant on 'laverock', from the Old English *láwerce*, literally 'treason worker', alluding to the parent lark's habit of deceiving intruders by the pretence of injury, and laying a limping false trail away from the nest. In the Orkney Islands it is 'Our Lady's hen' and associated with Virgin Mary.

The word 'lark' is also curiously close to the Middle English *laik*, meaning to play or fool around. Thus the nineteenth-century expressions 'larking about' and 'skylarking' may stem from the acrobatics of fighting skylarks or the jubilance of its singing, but they may separately stem from *laik* – which has also provided the *lek* of the black grouse and the *Leg* in the Danish plastic construction toy Lego®.

Common names for the skylark

Rising Lark, Sky Laverock (Northamptonshire)
Field Lark (Surrey)
Short-heeled Lark (Scotland)
Clod Lark, Clodhopper (Sussex)
Ground Lark, Sky Flapper (Somerset)
Laverock, Lavrock, Laverak (Scotland)
Learock (Lancashire)
Lerruck, Lavro (Orkney)
Linty white (Suffolk)
Melhuez (Cornwall)
Our Lady's Hen, Lady Hen (Shetland)

Skylark Miscellany

- *Alauda arvensis* is skylark, ground lark, but it is not water lark. The bird is near hydrophobic, and rarely if ever drinks, except in the artificial circumstances of captivity. Cleaning is by sunning and dust-bathing.

- The skylark is one of the most abundant and widespread of all the larks, breeding in a broad band across Eurasia, between about 40°N and 65°N, although there are regular breeding populations as far north as 72°N in Norway, well inside the Arctic Circle. In the extreme south of the species' range, birds have been recorded displaying in Qatar, although skylarks appear generally to be rare and erratic breeders on the Arabian Peninsula.

- 'Skylarking' was the name given to a naval practice where young sailors were encouraged to climb high in the rigging of sailing ships.

- Julius Caesar raised a legion known as Legio quinta alaudae, after its lark-crested helmets. The legion was destroyed in AD 86 at the battle of Tapae.

- Although a bird of farmland, the skylark can be found in a variety of other habitats. Richmond Park in London has had a breeding population for many years.

- The skylark may have evolved on the Asian steppe, and the bird still reaches its highest densities there, 113 birds per square kilometre.

- Homesick Britons, desirous of hearing the skylark's song, successfully introduced the bird to Australia, New Zealand, Canada and the Hawaiian islands in the nineteenth century.

- In medieval Britain, eating skylarks was believed to cure ailments of the throat. In Belgium, parents gave their children skylark flesh to ensure they became God-fearing. In Italy, eating lark was a cure for liver disease.

- In Ireland, the skylark is a godly bird; according to legend, the skylark is sacred to Saint Brigid and wakes her to prayers each morning.

To a Lark Singing in Winter

Wing-winnowing lark with speckled breast
Has just shot up from nightly rest
To sing two minutes up the west
Then drop again
Heres some small straws about her nest
All hid from men

Thou farmers minstrel ever cheery
Though winters all about so dreary
I dare say thou sat warm and erie
Between the furrows
And now thy song that flows unweary
Scorns earthly sorrows

The little mouse comes out and nibbles
The small weed in the ground of stubbles
Where thou lark sat and slept from troubles
Amid the storm
The stubbles ic'el began to dribble
In sunshine warm

Sweet minstrel of the farm and plough
When ploughmans fingers gin to glow
How beautiful and sweet art thou
Above his head
The stubble field is in a glow
All else seems dead

All dead without the stubble ground
Without a sight without a sound
But music sunshines all around
Beneath thy song
Winter seems softened at thy sound
Nor nips to wrong

On all the stubble-blades of grass
The melted drops turn beads of glass
Rime feathers upon all we pass
Everywhere sings
And brown and green all hues that was
Feathered like wings

It is a morn of ragged rime
The coldest blast of winter time
Is warmth to this Siberian clime
Dead winter sere
And yet that clod brown bird sublime
Sings loud and clear

The red round sun looks like a cheat
He only shines blood freezing heat
And yet this merry birds night seat
Seems warm's a sty
The stubble woods around it meet
And keep it dry

Each stubble stalk a jiant tree
Scarce higher than an infants knee
Seem woods to stop the winds so blea
From this snug home
Boundless this stubble wood must be

How safe must be this birds sweet bed
In stubble fields with storms o'er head
Or skies like bluest curtains spread
Lying so lone
With bit of thurrow o'er her head
Mayhap a stone

The god of nature guides her well
To choose best dwellings for her sell
And in the spring her nest will tell
Her choice at least
For God loves little larks as well
As man or beast

Thou little bird thou bonny charm
Of every field and every farm
In every season cold and warm
Thou sing'st thy song
I wish thy russet self no harm
Nor any wrong

Free from the snares thy nature shuns
And nets and baits and pointed guns
Dangers thy timid nature shuns
May thou go free
Sweet bird as summer onward runs
I'll list' to thee

I'd writ one verse, and half another
When thou dropt down and joined a brother
And o'er the stubble swopt together
To play 'till dark
Then in thy night nest shun cold weather
As snug's a Lark

Old russet fern I wish thee well
Till next years spring comes by itsel
Then build thy nest and hide it well
'Tween rig or thurrow
No doubt may be this is the dell
—Spring comes the morrow

Then blossomed beans will bloom above thee
And bumble bee buz in and love thee
And nothing from thy nest shall moove thee
When may shines warm
And thy first minstrel[sy] above thee
Sing o'er the farm

John Clare

II

The Lark in Music
and Poetry

S UCH IS THE skylark's popularity with poets that
the bird's biography could be told in verse. Of the
British birds, only the nightingale has been more
poeticized. In C. H. Poole's classic *Treasury of Bird
Poems* (1911), there are thirty-five skylark poems,
marginally less than the nightingale's thirty-nine.

As with the nightingale, skylark poetry is largely
about the bird's song. Unlike the nightingale, however, the skylark is not baggaged with legend. The
nightingale is Philomel, the maiden violated, her
tongue cut out so she could not report the crime, and
who is finally transformed into a bird. The bird's Latin
name *Luscinia* is from *luctus*, meaning 'lamentation',
and *canere*, 'to sing'. Melancholia was the sentiment
aroused in poets by the evensong of the nightingale.

The skylark, from the outset, had a happy time

with verse-makers. How could it be otherwise? Any knowledge of the bird, especially its singing habit, made it a natural motif for merriment. Early to the page was John Lyly (1553–1606), with the stanza:

> *Who is't now we hear?*
> *None but the lark so shrill and clear;*
> *Now at heaven's gate she claps her wings,*
> *The morn not waking till she sings.*

'Spring's Welcome'

The morn. A standard of skylark poetry would be the bird's ending of the night, with all its terrors and fears. Then again, the bird's scientific name *Alauda* is derived from the Latin for dawn, that is *aurora*.

Lark Lyly's near, but rather more luminous, contemporary William Shakespeare makes the bird a rival to Chanticleer in the honour of starting the day:

> *Lo, here the gentle lark, weary of rest,*
> *From his moist cabinet mounts up on high,*
> *And wakes the morning, from whose silver breast*
> *The sun ariseth in his majesty.*

Venus and Adonis

And:

> *The busy day,*
> *Waked by the lark, hath roused the ribald crows.*

Troilus and Cressida, IV, 2

The joyous sound of the bird's carol is commemorated in the line: 'And merry larks are ploughmen's clocks' in *Love's Labour's Lost*. And in *Cymbeline*: 'Hark, hark! the lark at heaven's gate sings'.

The bird-melodies of night and morning are commingled in the bedroom scene in *Romeo and Juliet*, where Juliet, from her window above, would try to persuade her lingering lover that it was not yet near day:

> *JULIET: Wilt thou be gone? it is not yet near day:*
> *It was the nightingale, and not the lark,*
> *That pierced the fearful hollow of thine ear;*
> *Nightly she sings on yon pomegranate-tree:*
> *Believe me, love, it was the nightingale.*
> *ROMEO: It was the lark, the herald of the morn,*
> *No nightingale: look, love, what envious streaks*
> *Do lace the severing clouds in yonder east . . .*

Romeo and Juliet, III, 5

Shakespeare's most condensed skylark moment comes in Sonnet 29, where the bird is the absolute antiphonal solution to misery:

> Yet in these thoughts myself almost despising,
> Haply I think on thee, and then my state,
> Like to the lark at break of day arising
> From sullen earth sings hymns at heaven's gate;
> For thy sweet love remembered such wealth brings
> That then I scorn to change my state with kings.

Quite the birdman was the Bard. There is misconception that love of birds, love of nature is intrinsically modern and metropolitan. George Orwell felt obliged to dish out a rejoinder in 'Some Thoughts on the Common Toad', *Tribune*, 12 April 1946:

> This is often backed up by the statement that a love of Nature is a foible of urbanised people who have no notion what Nature is really like. Those who really have to deal with the soil, so it is argued, do not love the soil, and do not take the faintest interest in birds or flowers, except from a strictly utilitarian point of view. To love the country one must live in the town, merely taking an occasional week-end ramble at the warmer times of year.

This last idea is demonstrably false. Medieval literature, for instance, including the popular ballads, is full of an almost Georgian enthusiasm for Nature, and the art of agricultural peoples such as the Chinese and Japanese centre always round trees, birds, flowers, rivers, mountains.

Orwell might have added Shakespeare to his list of enthusiasts for nature.

The singing lark of John Milton (1608–74) in 'L'Allegro' does 'startle the dull night'; the image of the bird stationed 'in a watch-towre in the skies' is unforgettable (and inspirational: Samuel Palmer etched in homage) but pure romantic fancy.

By the nineteenth century and the time of the Romantic poets proper, skylark poetry was as abundant as the bird. William Wordsworth, with the enthusiasm characteristic of the Romantics, declared in 'A Morning Exercise':

> . . . ne'er could Fancy bend the buoyant Lark
> To melancholy service – hark! O hark!

The skylark would often be attended by exclamation marks in Romantic verse. Wordsworth's skylark is the emblematic Romantic skylark, never sad, always merry.

Constitutionally, continuously cheery. Wordsworth even goes so far as to claim the skylark as 'The happiest bird that sprang out of the Ark!'

Cannot skylarks be angry or anxious? There is no room in Wordsworth for the skylark red in claw and beak from scrapping with rivals. Contrast Wordsworth with this cock-fight scene from D'Urban and Mathew's *The Birds of Devon* (1892): 'We had a new revelation respecting the character of the Sky-Lark when we were recently in a bird-catcher's shop, and in a long cage containing a dozen or more lately captured larks witnessed a most desperate combat between two young cocks, while the others stood round ruffled and bleeding from recent contests.'

In the wild, fighting skylarks fly up in the air, claws out, beat each other with wings, chase. All while singing. If skylarks are divine, they are also little hot balls of feather and bone. Birds with dual personality.

The most famous skylark poem of all, Percy Shelley's 'To A Skylark' – an exclamation mark in the very first sentence: 'Hail to thee, blithe spirit!' – has nothing to do with British larks; indeed it has next to nothing to do with flesh-and-blood larks. Or as Shelley writes in line two: 'Bird thou never wert.'

This essential of poetry ornithology, this staple of Eng Lit GCSE, was composed on or around 22 June

1820 as Shelley and his wife, Mary, walked down to the sea near Leghorn, Italy. In her note to the poem in the 1872 edition of her late husband's *Collected Works* Mary Shelley remembers a particular summer evening: 'while wandering among the lanes whose myrtle hedges were the bowers of the fire-flies . . . we heard the carolling of the skylark which inspired one of the most beautiful of his poems'.

The real subject of 'To a Skylark' is Shelley himself, whose hamartia was his egocentrism. For Shelley, the lark symbolizes the ultimate poet, being one free of constraint. The poet as lark is explicit in the stanza:

> *Like a poet hidden*
> *In the light of thought,*
> *Singing hymns unbidden,*
> *Till the world is wrought*
> *To sympathy with hopes and fears it heeded not.*

The saving grace of the poem is Shelley's acknowledgement of the difficulty of pinning the bird's voice in verse; as the skylark goes further up into the sky, words become ever more doomed to insufficiency:

> *Higher still and higher*
> *From the earth thou springest*

Like a cloud of fire;
The blue deep thou wingest,
And singing still dost soar, and soaring ever singest.

John Clare overlapped with the Romantics in era, and in the intensity of love for Nature. But he was more observant of the avian realities. Son of a farm worker, he lay in the fields of Helpston, as boy and man, among the larks. He knew that the skylark was also the ground lark, and:

That birds which flew so high would drop again
To nest upon the ground where anything
May come at to destroy.

'The Skylark'

Clare, so malnourished in childhood that he remained always a child in height (barely reaching five feet by his twenties), had an obsession with nests, sanctuaries, safe spaces. Home. He once proposed a collection entitled 'Birds Nesting'. So attached to the Helpston fields was Clare that when his patron, Earl Fitzwilliam, moved him and his family to a larger cottage in North-borough, just three miles along the lane, it provoked the mental crisis that preceded the poet's fall into drinking and madness. Clare wrote eight poems on skylarks in his lifetime.

Then there is George Meredith (1828–1909), infamous for his purply comi-tragic novels *The Ordeal of Richard Feverel* and *The Egoist*; time has been gentler on his poems, particularly the 122 lines of rhyming tetrameter that comprise 'The Lark Ascending'. This first appeared in *The Fortnightly Review* for May 1881, and has been lauded ever since. Siegfried Sassoon – a birdwatcher as well as a poet – thought it peerless, 'a sustained lyric which never for a moment falls short of the effect aimed at, soars up and up with the song it imitates, and unites inspired spontaneity with a demonstration of effortless technical ingenuity'. He might have added that Meredith's verse also captured, as well as the 'hurried press of notes', the uplifting effect of the bird's chansoning on us poor humans, akin to a fountain of light piercing the 'shining tops of day'. Such was the pull of the skylark, in truth and in culture, that even a declared atheist like Meredith has his skylark poem veer within a feather's blade of pantheism; in the lark's song human 'millions rejoice/For giving their one spirit voice'. In the poetry of the skylark there arose a distinct strain of religiosity.

The appeal of Meredith's 'The Lark Ascending' to such a committed pastoralist as Ralph Vaughan

Williams is no surprise. Now more widely known than the poem, Vaughan Williams's musical composition of the same title was written in August 1914 as he walked the Margate cliffs, watching naval vessels on manoeuvres. With the help of the English violinist Marie Hall, Vaughan Williams re-scored the piece for solo violin and orchestra, which better still suggested the way the lark's song spreads from its high source, filling the air with a progression of fine, tinkling notes. It is this version which has entered the collective head and heart of classical music lovers. In the Classic FM annual Hall of Fame poll, *The Lark Ascending* was placed first in 2014, 2015, 2016 and 2017. In 2011, in a radio poll of New Yorkers for preferences of music to commemorate the tenth anniversary of the 9/11 terrorist attacks, *The Lark Ascending* ranked second.

In retrospect, Vaughan Williams's *The Lark Ascending* seems a predictive requiem for pale-faced battalions in the Great War's trenches. The skylark flew there too.

A List of Skylark Poems

An Easter Legend of the Skylark, Florence Coates
The Skylark, John Clare
The Skylark (another version), —
The Skylark (another version), —
The Skylark's Nest, —
To a Skylark Singing in Winter, —
To the Skylark, —
Address to a Lark Singing in Winter, —
Larks and Spring, —
The Skylark, J. Conder
Aubade, Sir William Davenant
The Bunch of Larks, Robert Leighton
The Ecstatic, C. Day-Lewis
To a Skylark Behind Our Trenches, Edward de Stein
Skylark, Cicely Fox Smith
Hark, Hark the Lark, Ivor Gurney
Shelley's Skylark, Thomas Hardy
A Late Lark Twitters from the Quiet Skies,
    W. E. Henley
The Skylark, James Hogg
The Sea and the Skylark, Gerard Manley Hopkins
Skylark, Ted Hughes
Song of the Captive Lark, John Logan

The Lark in London, Gerard Massey
The Lark Ascending, George Meredith
To a Lark, F. W. Orde Warde
The Skylark, William Renton
Returning, We Hear the Larks, Isaac Rosenberg
The Skylark, Christina Rossetti
The Skylark, Percy Bysshe Shelley
The Skylark, William Shenstone
Shelley in the Trenches, John William Streets
A Lark above the Trenches, —
To the Lark, Robert Southey
Nesting (The Lark), W. Moy Thomas
The Quiet Singer, Francis Thompson
The Wounded Lark, Edwin Waugh
To Alauda, The Herald of Spring, W. Percival
    Westell
To a Skylark, William Wordsworth
To the Skylark, —

A Lark Above the Trenches

Hushed is the shriek of hurtling shells: and hark!
Somewhere within that bit of soft blue sky –
Grand in his loneliness, his ecstasy,
His lyric wild and free – carols a lark:

I in the trench, he lost in heaven afar,
I dream of Love, its ecstasy he sings;
Doth lure my soul to love till like a star
It flashes into Life: O tireless wings

That beat love's message into melody –
A song that touches in this place remote
Gladness supreme in its undying note
And stirs to life the soul of memory –
'Tis strange that while you're beating into life
Men here below are plunged in sanguine strife!

Sergeant John William Streets,
12/York and Lancaster Regiment

To a Skylark Behind Our Trenches

Thou little voice! Thou happy sprite,
How didst thou gain the air and light –
That sing'st so merrily?
How could such little wings
Give thee thy freedom from these dense
And fetid tombs – these burrows whence
We peer like frightened things?
In the free sky
Thou sail'st while here we crawl and creep
And fight and sleep
And die.

How canst thou sing while Nature lies
Bleeding and torn beneath thine eyes,
And the foul breath
Of rank decay hangs like a shroud
Over the fields the shell hath ploughed?
How canst thou sing, so gay and glad,
Whilst all the heavens are filled with death
And all the world is mad?

Yet sing! For at thy song
The tall trees stand up straight and strong
And stretch their twisted arms.
And smoke ascends from pleasant farms
And the shy flowers their odours give.
Once more the riven pastures smile,
And for a while
We live.

Captain Edward de Stein,
Machine Gun Corps and King's Royal Rifle Corps

III

The Skylarks of the Western Front

I T WAS THE conceit of the Edwardian British that the skylark was their bird. They knew it from school in the poetry of Shelley, Meredith, Wordsworth, and they knew it from a countryside bright with birdsong.

When the British in uniform arrived on the Western Front in 1914 for the first war against German militarism, they found their bird there before them. And miraculously it refused to budge. Sergeant H. H. Munro (better known as Scottish short-story writer 'Saki') noted how the skylark had 'stuck tenaciously' to the land despite the trenches, bullets and shells. He added:

It seemed scarcely possible that the bird could carry its insouciance to the length of attempting to rear a brood in that desolate wreckage of shattered clods and gaping shell-holes, but once, having occasion to throw myself

51

down with some abruptness on my face, I found myself
nearly on the top of a brood of young larks. Two of them
had already been hit by something, and were in rather a
battered condition, but the survivors seemed as tranquil
and comfortable as the average nestling.

'Birds on the Western Front'

The skylark's refusal ('brave' was the adjective usually attached) to quit its habitat because of warring man caused widespread admiration. The bird even stayed put on day one of the Somme, the never-to-be-forgotten 1 July 1916, the bloodiest day in British military history, with its 58,000 British casualties. The attack was launched on a thirty-kilometre front, between Arras and Albert.

Despite the storm of shot and shell, the skylarks rose into the cerulean sky (Siegfried Sassoon, at Fricourt with the Royal Welch Fusiliers, called the weather 'heavenly') to sing. The correspondent for *The Times* informed readers that the skylarks could be heard singing during the battle 'whenever there was a lull in the almost incessant fire'. A month later, with the battle still raging, Sergeant Leslie Coulson, 1/12 London Regiment, in the line at Hébuterne, paid his admiring respects to the magnificent tenacity of the Somme's *Alauda arvensis* in poetry, which was the soldier's medium:

From death that hurtles by
I crouch in the trench day-long,
But up to a cloudless sky
From the ground where our dead men lie
A brown lark soars in song.
Through the tortured air,
Rent by the shrapnel's flare,
Over the troubleless dead he carols his fill,
And I thank the gods that the birds are
     beautiful still.

'The Rainbow'

Coulson was killed charging the German line at Le
Transloy on the Somme in October 1916.

Second Lieutenant Otto Murray-Dixon, Sea-
forth Highlanders, was amazed during a German
'hate' or shelling that: 'The larks continue to sing
right through the shelling and kestrels hover about,
not at all concerned by all the noise; it is very cheering
to see them.' Murray-Dixon had spent his childhood
rambling and bird-nesting around his home, the Old
Rectory at Swithland in Leicestershire, prior to
attending Calderon's School of Animal Painting.
Before joining up, he had been commissioned by J. G.
Millais to provide drawings for his British Diving

Ducks. One of the obituaries for Murray-Dixon, following his death in battle in 1917, observed 'wild flowers filled his soul'.

The skylarks were there in the trenches of Flanders too. Poppies might dominate the landscape of Lieutenant-Colonel John McCrae's iconic 'In Flanders Fields', but there's nonetheless a significant place for the skylark:

> *In Flanders fields the poppies blow*
> *Between the crosses, row on row,*
> *That mark our place; and in the sky*
> *The larks, still bravely singing, fly*
> *Scarce heard amid the guns below.*

The song of the skylark was the musical background to the war on the Western Front, as it was to life in the British countryside. And it was clock, and it was calendar. For Private Norman Edwards, 6/Gloucestershire Regiment and in the line at Plugstreet Wood in May 1915, the melody of the skylark was part of the pattern of the soldier's day:

> *One realised how close one was living to nature, closer perhaps than ever before, and the thought that possibly each dawn might be the last accentuated the delight. The*

dawns at this time were particularly beautiful. Before any definite light appeared, the larks would soar up and a faint twittering in the wood grew to a buzz of noise as the birds stood-to with us.

'The song of the Skylark at dawn over No-Man's-Land was as usual as the song of the sniper's bullet,' believed the *Daily News and Leader*. The matins of the skylark was more than an alarm clock, it was confirmation that the black night – when all manner of dangerous tasks, from patrolling to fetching supplies, were undertaken – was over.

Private Isaac Rosenberg and his mates, coming back to camp after a night stint of trench maintenance, were delighted to hear larks in the sky rather than the expected whine of bombs. It was an interlude from death, a delight shared by all men, with the larks themselves forming an invisible shield:

Sombre the night is:
And, though we have our lives, we know
What sinister threat lurks there.

Dragging these anguished limbs, we only know
This poison-blasted track opens on our camp—
On a little safe sleep.

But hark! Joy—joy—strange joy.
Lo! Heights of night ringing with unseen larks:
Music showering on our upturned listening faces.

Death could drop from the dark
As easily as song—
But song only dropped,
Like a blind man's dreams on the sand
By dangerous tides;
Like a girl's dark hair, for she dreams no ruin lies there,
Or her kisses where a serpent hides.

Written on scraps of paper, sent by Rosenberg to his parents, 'Returning, We Hear the Larks' was published in *Poetry* magazine after his death on 1 April 1918. A sniper's bullet, not song, dropped from the dawn sky that day.

In the unromantic environment of the trenches, the lark continued to be Romantic poetical inspiration. Sergeant John William Streets, 12/York and Lancaster Regiment ('Sheffield Pals'), explicitly acknowledged his debt to the Romantics' Romantic in 'Shelley in the Trenches':

A lark trill'd in the blue: and suddenly
Upon the wings of his immortal ode

My soul rushed singing to the ether sky
And found in visions, dreams, its real abode –
I fled with Shelley, with the lark afar,
Unto the realms where the eternal are.

Eldest of twelve children, unable to take up a place at grammar school due to poverty, Streets had worked on the coal face at the Whitwell mine, Derbyshire, but filled his spare time studying French and the Classics. The pit and the trench were both troglodyte worlds. Streets was killed on the first day of slaughter on the Somme, when the 'Sheffield Pals' went over the top at Zero Hour – 0730 hours – near Serre, to be met, in the words of the battalion's war diary, 'with very heavy machine gun and rifle fire and artillery barrage'.

Streets was aged thirty-one when he died. In May 1917 a collection of his war poems was published posthumously as *The Undying Splendour*, which was reasonable titling, both for the quality of the poetry and the manner in which Streets and his Pals met their deaths. Appreciation of the skylark, however, was not dependent on being groomed, or even self-educated, in culture. The musical song of the bird was intrinsically beautiful, and lifted spirits. Private Ivor Gurney (another poet of the lark), doing his rifle training, lying on his belly, reloading 'with the quickness of those who get extra pay for it', saw a skylark fly up. Stretched out next to him was Tim Godding, 'a Shakespearean character':

Now Tim Godding has little bits of jargon, some of which I strongly suspect to be Hindustani. One of these is 'ipsti pris', a sign of high spirits, of salutation to a passing battalion or the crown of a joke: anything joyful. So Tim Godding half turned over, looked up to the first blue of spring – 'Ipsti pris, skylark, ipsti pris'!

Skylarks turned the eyes upwards from present problems. When Siegfried Sassoon wrote in his diary on 3 June 1916 from near Sailly Laurette that 'A lark goes up, and takes my heart with him', it was nature pure providing the uplift rather than literary association.

Watching skylarks and other birds was a simple pleasure, an interest that did not have to be invested with meaning. 'Birding' was probably the most popular single hobby among troops on the front line. The Western Front was curiously good birdwatching country. After all, an enormous corridor of it, as much as a mile across, was composed of 'no man's land' where, as the name indicated, humans feared to tread. In a letter home to his wife Maude, Captain Charlie May of the Manchesters described the front in frosty March 1916:

Just over the brink of the parapet one catches a glimpse of jagged earthenware, the remains of a rum jar, or the battered lid, rising oyster fashion, from a discarded jam tin. That is No Man's Land, a portion of the earth where Tommy can with impunity gratify his natural tendency to untidiness by flinging to it all that endless rubbish which a battalion mysteriously accumulates even in the short span of a day. Only birds live out there – apparently as happily as ever. A lark trills blithefully somewhere up in the heavens above even as I write, his note throbbing as though 'twould burst his throat, full of the joy of the dawning and the promise of spring.

Birds were objects of interest, comfort, amusement to soldiers serving on the Western Front. When one soldier, a former Dunfermline miner, told his newspaper, 'What a hell it would be without the birds', he spoke for a generation in uniform. And of all the birds of the Western Front the skylark was the greatest solace to British soldiers.

The Caged Skylark

As a dare-gale skylark scanted in a dull cage
Man's mounting spirit in his bone-house,
    mean house, dwells—
That bird beyond the remembering his free fells;
This in drudgery, day-labouring-out life's age.
Though aloft on turf or perch or poor low stage,
Both sing sometímes the sweetest, sweetest spells,
Yet both droop deadly sómetimes in their cells
Or wring their barriers in bursts of fear or rage.

Not that the sweet-fowl, song-fowl, needs no rest—
Why, hear him, hear him babble & drop down to his nest,
But his own nest, wild nest, no prison.

Man's spirit will be flesh-bound when found at best,
But uncumberèd: meadow-down is not distressed
For a rainbow footing it nor he for his bónes rísen.

Gerard Manley Hopkins

II. *On the* Lark *and* Fowler.

IV

The Hunting of the Lark

Judging by the ordinary laws of probability the English Sky-lark ought long ago to have gained the distinction that belongs to the Dodo and the Great Auk, for the Lark abides in the midst of foes.

Society for the Protection of Birds,
Educational Series No. 22 – *Skylark* (1897)

How our attitude to the skylark has changed. Nowadays an object of devotion for nature-lovers, until the end of the nineteenth century the skylark was largely regarded as something for plate and cage.

To catch and kill larks, humans have developed all manner of techniques. Historically, the commonest method of catching skylarks was to flush the birds into a 'night-lark-net', or trammel, a net around twenty metres wide and braced by poles along each edge. The trammel was dropped on to the roosting birds. Skylarks, fatally, sleep on the ground. Trammelling required absolute darkness; on nights without moon or star-glow 400 birds or more could be caught by two men. A variation on the method was 'low-belling', when the net was preceded by men or boys ringing bells similar to those worn by cattle; the tolling of the bells caused the birds to remain on the ground until the net was over them. (The birds perhaps mistook the sound for real cows wearing bells, a sound known to them, and not alarming.) Low-belling was so highly valued in the sixteenth century that it was a right included – or not – in land tenure. Other methods for catching larks were drag netting, again at night, or in daylight employing 'double clap' nets, where two netting 'doors' were pulled down tight over a landed bird attracted by decoy.

The shotgun revolutionized the killing of larks for food. Birds, especially those on migration, were attracted to the guns by 'lark mirrors', a revolving device, thirty to fifty centimetres long, inlaid with pieces of looking glass or plain glass. When migrating skylarks were passing over the hunter, the device was spun rapidly, to extraordinary effect on the birds, as A. E. Knox recorded in his *Ornithological Rambles in Sussex* (1850):

> . . . *the reflection of the sun's rays from these little revolving mirrors seems to possess a mysterious attraction for the larks, for they descend in great numbers from*

a considerable height in the air, hover over the spot, and suffer themselves to be shot at repeatedly without attempting to leave the field or to continue course.

In France, lark mirrors were sufficiently commonplace to become the substance of an idiom, *'alouettes au miroir'*, meaning a strong but dangerous attraction. Like 'moths to a flame'. There are records of over 1,000 skylarks being shot at a single mirror on a single day in the nineteenth century.

Victorian Lark Recipes

LARKS (in Aspic).
Ingredients: 12 larks, 1 pint stock, 1 oz truffles, aspic.

Method: Put the larks in a small stewpan, cover them with the stock, and braise them carefully for 20 minutes, then set them aside to become quite cold. Melt the aspic, and put a tea-spoonful into each little mould, cut your truffles into tiny strips, and form a star at the bottom of each mould, put in the larks breast downwards, and with a spoon very carefully fill up the moulds with aspic, taking great care not to disarrange the truffles or to set the larks crookedly. When all are done, set them aside in a cold place for some hours until quite firm. To turn out, dip each mould for a moment into boiling-water, and turn them on to the dish exactly as you wish them placed, so as not to have to move them. Garnish with very fresh parsley. *Time*: To braise larks, 20 minutes. Sufficient for 5 or 6 persons.

LARK PUDDING.

Ingredients: 1 dozen larks, 1lb. steak, 1 oz. flour, ½ teaspoonful pepper and salt, crust No. 404, made with ¾ lb. flour, and other ingredients in proportion, ½ pint stock or water.

Method: Mix the flour, pepper, and salt together on a plate, cut the steak into neat pieces, and dip each lark and piece of steak into the mixture. Butter a pudding-basin, and line it with crust, pack the basin with the larks and steak, pour in the stock or water, and fit in the cover neatly, pinching the edges together. Tie it up in a floured cloth, plunge it into a saucepan of boiling water, and be careful not to let the water go off the boil until the pudding is cooked. Dish the pudding in the basin with a serviette neatly pinned round. *Time*: 4 hours. Sufficient for 5 or 6 persons.

From A *Practical Dictionary of Cookery*, Ethel Meyer, 1898

Larks are ubiquitous in recipe books from Roman times onwards. The Romans considered larks' tongues an epicurean delight, an *amuse-bouche*, whereas the medieval English regarded the whole bird as a staple food, as indicated by the low price demanded for them by the Worshipful Company of Poulters of the City of London. In the reign of Edward I the guild set the cost of a dozen skylarks as the equivalent of one mallard, or 1*d*. Prices of skylarks were also controlled in the other main European centre of lark consumption, Leipzig, 'whose Larks are famous all over Germany, as having the most delicate flavour'. The excise on the sale of Leipzig's larks raised about £60,000 in today's money each year in the late 1700s.

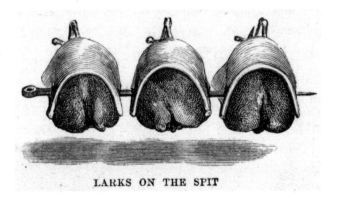

LARKS ON THE SPIT

John Ray in his collection *English Proverbs* (1673) has 'One leg of a Lark is worth the whole body of a kite', and 'the Land where Larks fall ready roasted' is heaven on earth. So the lark was regarded as good eating, as well as cheap meat.

How were larks eaten? Skylarks are tiny, tinier still when de-feathered and de-beaked. Roasting was the standard method of cooking. Afterwards each bird was popped in the mouth whole, in the way larks, ortolans and other small birds are still eaten in France and the Mediterranean countries. To enhance the experience – and I saw this as a child in a restaurant on Palma harbour in the 1970s – a cloth could be placed over the head when consuming the bird. French hunters still take a million or more larks and small birds per year.

The eating of larks in Britain reached its apogee in the Victorian century, when the lark became a culinary vogue among the rich, while still being consumed in appreciable numbers by the middle and lower classes. The Victorian lark was roasted, baked, put in a pie, entombed in aspic, put in pudding, turned into *mauviettes en surprise aux truffes* in St James's clubs. In 1854 the number of skylarks reaching the London meat markets, primarily Leadenhall, was 400,000 per annum. But the skylark trade had not yet reached its peak. By the 1890s as many as 40,000 skylarks were sold in the London markets per day. Hunting for these larks was carried out mostly in south-east England, particularly the Sussex coast, the Home Counties and Cambridgeshire. The larks of Dunstable Downs, Bedfordshire, were considered the finest, the skylark equivalent of Wye salmon, Arbroath kippers.

There were other markets for the bird-catchers, in other places. Over a million British larks went to the Dieppe market in 1867–8, and a similar amount to Les Halles, Paris.

Not all the birds taken by the bird-catchers ended up in pot or oven. There was a division of labour for the trapped skylark; the female went for meat, the cock went to the cage-bird trade. The clothing trade had its pound of feathers too; skylark wing feathers were dyed to imitate the feathers of exotic species.

Larks, like other singing birds, were prized as caged birds, and were blinded on the basis of a superstition that they sang better with their eyes put out. The Victorian traffic in songbirds was big business. A good singer of a skylark might change hands for fifty shillings; a bricklayer's pay for a six-day week was seven shillings. In cages skylarks still sang, as if the cage bottom were a meadow, and the cage top the vault of the

No. D 741.   LARKS' CAGES.
Painted Green.   Tinned Wire.

Small	Middle	Large.
9,	10,	11 inches,
1/11,	2/11,	3/4 each.

No. D 742.   LARKS' CAGES.
Painted Green.   Tinned Wire.
With Blank Side.

Small.	Middle.	Large.
9,	10,	11 inches.
2/6,	3/4,	3/10 each.

sky. No lesser person than the eminent ornithologist Henry Seebohm opined:

> No bird is better known or more frequently kept in confinement than the Sky-Lark. It is easily caught, readily tamed, and bears its captivity well, singing as sweetly on the sod in its little cage as when soaring in boundless freedom high up amongst the clouds.

The ability of skylarks to mimic other species was recognized by Victorian dealers of caged birds. In order to prevent their caged birds from losing their original song in favour of imitations of other captive birds, dealers placed them in cages next to newly caught birds to keep them, in the argot, 'honest'.

Birds awaiting sale were more often than not kept in appalling conditions. One eyewitness wrote in 1864:

Almost daily I pass a shop in Old-street, St Luke's, and there, exhibited in the open window, may always be seen a sight loathsome and sickening. Store-cage packed on store-cage, and each one literally crammed with birds of various kinds, larks, linnets, thrushes &c. To say the least, each cage is half-full of birds, so that they perch on each other's backs, while, at the same time, the cages themselves are as filthy and disgusting as can be imagined. I wonder what prevents the officers of the Society for the Prevention of Cruelty to Animals from giving the bird dealer in question a call.

The skylark trade aroused high passions in Victorian England, and saving the skylark became an early and totemic campaign of the Society for the Protection of Birds, founded in Didsbury in 1891, and granted a Royal Charter in 1904. The ornithologist William Henry Hudson (one of the few men in the Society) did not mince his words about lark-eating:

But the feeling of intense disgust and even abhorrence the practice arouses in all lovers of nature grows, and will continue to grow; and we may look forward to the time when the feeders on skylarks of to-day will be dead and themselves eaten by worms, and will have no successors in all these islands.

The tide of morality was on the side of the SPB. Love of nature was becoming a British condition, manifesting itself in everything from the new national hobby of gardening to the folk-influenced music of classical composers such as Ralph Vaughan Williams. (What did Vaughan Williams write in the first week of the Great War? *The Lark Ascending*, the very musical embodiment of pastoralism and patriotism, the same pastoralism and patriotism that would inspire him, despite being, at

forty-two, over age, to volunteer to serve.) The British led the world in the keeping of pets, animal welfare legislation and a regard for birds so marked by 1910 that *Punch* declared their feeding to be a national pastime, dockers and clerks of London included.

Something of this nature-love is explained by the torn-from-the-rural-womb early industrialization, which left a psychic wound in the mind of the new town-dwellers. The injury was so abysmal that urban Britons recreated the countryside in their back gardens; the lawn is nothing but the country meadow brought, as salve, into *urbia*. Charitably, one can see the keeping of cage-birds such as the skylark as the bringing of nature into the home.

Nature was not 'other', separate, a thing apart to the British a century ago. Transport, in the shape of the railways, bicycles and (for the rich) the car enabled Edwardian city dwellers to explore the countryside. And what they found was a place of bottomless peace and bountiful nature. And the song of birds, of the skylark especially.

An odd celebrity proof of the wonder of the countryside in the first decades of the twentieth century occurred on 9 June 1910 in the little village of Itchen Abbas in Hampshire, when the British Foreign Secretary, Sir Edward Grey, and former US President

Theodore Roosevelt went for a 'bird walk'. Sir Edward Grey is known to everyone because of a single, immortal quotation. Looking out of his office window at dusk on 3 August 1914 and towards St James's Park, he remarked to his friend J. A. Spender: 'The lamps are going out all over Europe. We shall not see them lit again in our lifetime.' Grey was always looking out of the window. He was a devout ornithologist, hence his invitation to Roosevelt to accompany him on a twelve-mile walk around the Itchen Valley. Grey recalled that they were 'lost to the world' for hours. They saw forty separate species of bird, including, of course, the skylark. Grey later wrote *The Charm of Birds*, first published in 1927, the core message of which was that watching and listening to birds could bring solace and regeneration to the world-weary. It is still a valid message.

'The SPB's skylark campaign was two-pronged, pressuring Parliament for legislation, and organizing public protest. Members wrote to shops and markets, threatening to withdraw their custom unless skylarks were removed from sale. The society also commissioned scientific research – among the first ever on behalf of a conservation organization – which proved that only 13 per cent of the skylark's diet was injurious to agriculture. In *Birds Useful and Harmful*, published in 1909 and a fixture on the shelves of middle-class

homes (including this author's paternal grandparents),
Otto Herman and J. A. Owen were able to state:

In England larks have been very largely eaten, but happily
the practice is now most strongly opposed by thoughtful
people. If the consumption of Larks in our country went
on as it was doing a few years ago the species would soon
be extinct.

Issues relating to the legal protection of the skylark rumbled on until the 1930s. Although eating and
caging skylarks had all but stopped due to the cam-
paigning of the SPB (from 1904, the Royal Society for
the Protection of Birds), Section 6 of the Wild Birds
Protection Act 1931 allowed the killing of the skylark
by farmers. The vein of belief that the skylark was the
farmer's foe had persisted, despite the RSPB's evidence
to the contrary.

Full legal protection for the skylark came only
with the Protection of Birds Act 1954. But by then, a
threat far more dangerous than hunting and caging
faced the skylark. The new Agricultural Revolution
had begun.

To the Skylark

Ethereal minstrel! pilgrim of the sky!
Dost thou despise the earth where cares abound?
Or, while the wings aspire, are heart and eye
Both with thy nest upon the dewy ground?
Thy nest which thou canst drop into at will,
Those quivering wings composed, that music still!

Leave to the nightingale her shady wood;
A privacy of glorious light is thine;
Whence thou dost pour upon the world a flood
Of harmony, with instinct more divine;
Type of the wise who soar, but never roam;
True to the kindred points of Heaven and Home!

William Wordsworth

V

SOS: Save Our Skylarks – The Lark, Farming and Conservation

The hamlet stood on a gentle rise in the flat, wheat-growing north-east corner of Oxfordshire. We will call it Lark Rise because of the great number of skylarks which made the surrounding fields their springboard and nested on the bare earth between the rows of green corn.

Flora Thompson,
Lark Rise to Candleford (1939)

You would be hard-pressed to find a corner of Oxfordshire, or any other agricultural county in Britain, where larks today rise in 'great number' such as Thompson knew in the 1930s. The collapse of the skylark population – in the last twenty-five years alone we have 'lost' something like a million and a half pairs of skylarks in the UK – coincides precisely with the rise of intensive agriculture.

The modern agricultural revolution has its polit-ical roots in the Second World War, when most European countries faced crippling food shortages that during peacetime were made up by imports from abroad. After the war, the same countries saw self-sufficiency in agricultural produce as a strategic priority, to be solved by the creation of a 'common market' based on the removal of protective tariffs to member nations of the European Economic Community, the EEC (later the European Union). The EEC's Common Agricultural Policy protected producers by guaranteeing fixed prices, and by imposing tariffs on cheaper imported produce.

This artificially induced financial security proved the basis of intensification, since, for the first time, higher yields meant higher incomes. Guaranteed. The CAP also provided capital grants to give farmers access to the machinery needed to increase output. Result? The most rapid intensification of farming methods ever seen. Agriculture in the EU became geared almost exclusively to productivity, indeed to over-productivity, hence wine lakes and grain mountains.

The skylark in Britain is a farmland bird. A survey by the British Trust for Ornithology suggests that over 70 per cent of all the UK's skylarks are found on lowland farmland. Consequently any changes in the farming regime hit the bird hard.

On grassland, intensification has led to increased stocking densities of cattle and sheep on grazing land, meaning grass is chewed too short, too often for skylarks to nest, while the risk of nests being trampled by hooves is increased. Then there is the big switch from hay to silage for winter fodder, which sees nests, chicks, fledglings destroyed by the cutting machinery three or four times over the breeding season. Mowing not only destroys nests, it also makes the grass too short to provide cover for replacement nesting attempts. In one experiment in southern England, mowing of grass caused numbers of breeding skylarks to decline by around 80 per cent. In traditional hay systems, only a single cut is taken at the end of the summer, by which time the birds have largely finished breeding. More, the 'sward' for silage tends to be low in the number of grasses and flowers ('weeds') it contains, and may hardly set seed. It might also be sprayed with pesticide, doused with inorganic fertilizer, reducing invertebrate diversity. So there is neither home nor food for skylarks. Over 90 per cent of British grassland has been 'improved' or 'semi-improved' during the twentieth century. Organic and traditional grassland has up to five times the number of skylarks as treated grass.

The picture for skylarks on arable land, which is their preferred farmland habitat, is no better. Modern cereal fields for wheat, barley and oats are often too densely sown for skylarks to nest in. The big switch from spring to autumn sowing also means that the crop is the wrong height in the nesting season. The skylark is the Goldilocks bird when it comes to nesting. It doesn't like herbage to be too tall (the bird cannot see the nest, or predators) or too short (no camouflage given). Best is herbage between five and forty centimetres tall. There is some evidence that modern cereal farming, with its earlier harvest (which again subjects nests to a whirl of steel blades) is causing birds to cease all laying as early as May.

It gets worse. In dense cereal crops many nesting attempts are on or close to tramlines (tractor tracks that are used to apply the many sprays to the crop), which makes the nests vulnerable to ground predators. And the tractor's wheels.

It gets yet worse. Wintering skylarks are keen on overwinter cereal 'stubble', with its spare grains and protective stalks, left after the harvest cut. But only 3 per cent of UK farmland is now left as stubble over winter.

Can anything be done? Yes, and easily. One simple solution for farmers is the 'skylark plot', which costs next to nothing, in money or time. Research carried out by the RSPB shows small areas, 2 × 16–24 metres square per hectare, left unsown in winter cereals boost the nesting opportunities and food availability for skylarks. Skylark plots help the birds forage for food once the crops grow and become dense, by allowing the skylarks to land in the field and access insects in the crop. To create the patch, all that is needed is for the seed drill to be switched off for a minute.

According to the RSPB, skylark plots have increased the numbers of skylarks at the charity's Hope Farm in Cambridgeshire by nearly 50 per cent. In a decade.

If there were a thousand such plots, the decline of the skylark could be halted.

Beneficial Management for Skylarks on Farmland

- Include a spring cereal as part of the arable rotation. This provides ideal and much needed late-season nesting habitat.

- Retain overwinter stubble, especially cereal stubbles, to provide a source of winter food and a nesting habitat in spring/summer.

- Buffer strips and field margins will provide similar conditions and can attract very high breeding densities. Aim for a range of grass heights and structures.

- Be mindful of nesting birds and fledglings when cutting silage, fallow and buffer strips.

- Consider including 'Skylark plots' within winter cereals.

Game & Wildlife Conservation Trust

The Decline of Britain's Farmland Birds

The skylark, now 'red listed' as a target for conservation action, is only one of a number of farmland birds which have seen plummets in population. Don't take my word for it. Take that of the Department of the Environment, Food and Rural Affairs, publisher of 'Wild Bird Populations in the UK, 1970 to 2014', regarding farmland birds:

- *In 2014, the breeding farmland bird index in the UK was less than half (a decline of 54 per cent) of its 1970 level – the second lowest level recorded.*

- *Within the index over the long-term period, 21 per cent of species showed a weak increase, 21 per cent showed no change and 58 per cent showed either a weak or a strong decline.*

- *Most of the decline for the farmland bird index occurred between the late seventies and the early nineties, largely due to the impact of rapid changes in farmland management during this period.*

- *The smoothed indicator shows a significant on-going decline of 11 per cent between 2008 and 2013.*

  The farmland bird index comprises 19 species of bird. The long term decline of farmland birds in the UK has been driven mainly by the decline of those species that are restricted to, or highly dependent on, farmland habitats (the 'specialists') . . . Between 1970 and 2014, populations of farmland specialists declined by 69 per cent while farmland generalist populations declined by 9 per cent . . . Overall, 75 per cent of the 12 specialist species in the farmland indicator have declined.

The farmland 'specialist' birds are: corn bunting, goldfinch, grey partridge, lapwing, starling, stock dove, tree sparrow, turtle dove, white throat, linnet. And the skylark.

The Lark Ascending

He rises and begins to round,
He drops the silver chain of sound
Of many links without a break,
In chirrup, whistle, slur and shake,
All intervolv'd and spreading wide,
Like water-dimples down a tide
Where ripple ripple overcurls
And eddy into eddy whirls;
A press of hurried notes that run
So fleet they scarce are more than one,
Yet changingly the trills repeat
And linger ringing while they fleet,
Sweet to the quick o' the ear, and dear
To her beyond the handmaid ear,
Who sits beside our inner springs,
Too often dry for this he brings,
Which seems the very jet of earth
At sight of sun, her music's mirth,
As up he wings the spiral stair,

A song of light, and pierces air
With fountain ardor, fountain play,
To reach the shining tops of day,
And drink in everything discern'd
An ecstasy to music turn'd,
Impell'd by what his happy bill
Disperses; drinking, showering still,
Unthinking save that he may give
His voice the outlet, there to live
Renew'd in endless notes of glee,
So thirsty of his voice is he,
For all to hear and all to know
That he is joy, awake, aglow,
The tumult of the heart to hear
Through pureness filter'd crystal-clear,
And know the pleasure sprinkled bright
By simple singing of delight,
Shrill, irreflective, unrestrain'd,
Rapt, ringing, on the jet sustain'd
Without a break, without a fall,
Sweet-silvery, sheer lyrical,
Perennial, quavering up the chord
Like myriad dews of sunny sward
That trembling into fulness shine,
And sparkle dropping argentine;
Such wooing as the ear receives

SOS: Save Our Skylarks

From zephyr caught in choric leaves
Of aspens when their chattering net
Is flush'd to white with shivers wet;
And such the water-spirit's chime
On mountain heights in morning's prime,
Too freshly sweet to seem excess,
Too animate to need a stress;
But wider over many heads
The starry voice ascending spreads,
Awakening, as it waxes thin,
The best in us to him akin;
And every face to watch him rais'd,
Puts on the light of children prais'd,
So rich our human pleasure ripes
When sweetness on sincereness pipes,
Though nought be promis'd from the seas,
But only a soft-ruffling breeze
Sweep glittering on a still content,
Serenity in ravishment.

For singing till his heaven fills,
'T is love of earth that he instils,
And ever winging up and up,
Our valley is his golden cup,
And he the wine which overflows
To lift us with him as he goes:

The woods and brooks, the sheep and kine
He is, the hills, the human line,
The meadows green, the fallows brown,
The dreams of labor in the town;
He sings the sap, the quicken'd veins;
The wedding song of sun and rains
He is, the dance of children, thanks
Of sowers, shout of primrose-banks,
And eye of violets while they breathe;
All these the circling song will wreathe,
And you shall hear the herb and tree,
The better heart of men shall see,
Shall feel celestially, as long
As you crave nothing save the song.
Was never voice of ours could say
Our inmost in the sweetest way,
Like yonder voice aloft, and link
All hearers in the song they drink:
Our wisdom speaks from failing blood,
Our passion is too full in flood,
We want the key of his wild note
Of truthful in a tuneful throat,
The song seraphically free
Of taint of personality,
So pure that it salutes the suns

SOS: Save Our Skylarks

The voice of one for millions,
In whom the millions rejoice
For giving their one spirit voice.

Yet men have we, whom we revere,
Now names, and men still housing here,
Whose lives, by many a battle-dint
Defaced, and grinding wheels on flint,
Yield substance, though they sing not, sweet
For song our highest heaven to greet:
Whom heavenly singing gives us new,
Enspheres them brilliant in our blue,
From firmest base to farthest leap,
Because their love of Earth is deep,
And they are warriors in accord
With life to serve and pass reward,
So touching purest and so heard
In the brain's reflex of yon bird;
Wherefore their soul in me, or mine,
Through self-forgetfulness divine,
In them, that song aloft maintains,
To fill the sky and thrill the plains
With showerings drawn from human stores,
As he to silence nearer soars,
Extends the world at wings and dome,

More spacious making more our home,
Till lost on his aërial rings
In light, and then the fancy sings.

George Meredith

USEFUL ORGANIZATIONS

UK Biodiversity Action Plan: www.ukbap.org.uk
English Nature: www.english-nature.org.uk
The Royal Society for the Protection of Birds:
    www.rspb.org.uk
British Trust for Ornithology: www.bto.org
Game & Wildlife Conservation Trust:
    www.gwct.org.uk
Countryside Restoration Trust:
    www.countrysiderestorationtrust.com

PICTURE CREDITS

Other books in this series:

THE SECRET LIFE OF THE OWL

There is something about owls. They are creatures of the night, and thus of magic. But – with the sapient flatness of their faces, their big, round eyes, their paternal expressions – they are also reassuringly familiar. We see them as wise, like Athena's owl, and loyal, like Hedwig. Here, John Lewis-Stempel explores the legends and history of the owl. And in vivid, lyrical prose, he celebrates all the realities of this magnificent creature, whose natural powers are as fantastic as any myth.

THE PRIVATE LIFE OF THE HARE

The hare is a rare sight for most people. We know them only from legends and stories. They are shape-shifters, witches' familiars and symbols of fertility. They are arrogant, as in Aesop's *The Hare and the Tortoise*, and absurd, as in Lewis Carroll's Mad March Hare. In the absence of observed facts, speculation and fantasy have flourished. But real hares? What are they like? In elegant prose John Lewis-Stempel celebrates how, in an age when television cameras have revealed so much in our landscape, the hare remains as elusive and magical as ever.

THE GLORIOUS LIFE OF THE OAK

The oak is our most beloved and most familiar tree.
For centuries oak touched every part of a Briton's life –
from cradle to coffin. It was oak that made the 'wooden
walls' of Nelson's navy, and the navy that allowed
Britain to rule the world. John Lewis-Stempel explores
our long relationship with this iconic tree and retells
oak stories from folklore, myth and legend – oaks
bearing the souls of the dead, the Green Man and
fertility rites on Oak Apple Day. Of all the trees,
it is the oak that speaks most clearly to us.

THE WILD LIFE OF THE FOX

The fox is our apex predator, our most beautiful and
clever killer. We have witnessed its wild touch, watched
it slink by bins at night and been chilled by its high-
pitched scream. And yet we long to stroke the tumbling
cubs outside their tunnel homes and watch the vixen
stalk the cornfield. Foxes captivate us like no other
species. Exploring a long and sometimes complicated
relationship, *The Wild Life of the Fox* captures our
love – and sometimes loathing – of this magnificent
creature in vivid detail and lyrical prose.